The R.A.M.S. Library of Alchemy

Volume 21

Alchemical Symbols

by

Philip N. Wheeler

R.A.M.S. Publishing Company

Alchemical Symbols

by

Philip N. Wheeler

Produced by

Restorers of Alchemical Manuscripts Society

R.A.M.S. Publishing Company

R.A.M.S. Publishing Company
7309 East 102nd Street
Kansas City Missouri 64134

The R.A.M.S. Library of Alchemy, Volume 21:
Alchemical Symbols
Copyright © 2018 Althea Productions LLC

First Edition 2007
Second Edition 2011
Third Edition 2015
Fourth Edition 2018

ISBN-13 **978-1723544491**
ISBN-10 **1723544493**

Image Processing by Philip N. Wheeler

Printed in the United States of America

Table of Contents

Alchemical Symbols

Dedicated to Hans W. Nintzel,

American Alchemist

and

Founder of the

Restorers of Alchemical Manuscripts Society

(R.A.M.S.)

Disclaimer

Liability: The publisher does not warrant or assume any legal liability or responsibility for the accuracy, completeness, or usefulness of any information, apparatus, product, or process disclosed. The publisher makes no representation as to the accuracy or completeness of the contents of this book and specifically disclaims any implied warranty of merchantability or fitness for a particular purpose. No warranty may be created or extended by written sales materials or sales representatives. You should obtain professional consultation where appropriate. The publisher shall not be liable for any loss of profit or other commercial or personal damages, including but not limited to special, incidental, consequential, or other damages.

This book is sold for informational purposes only. Neither the publisher nor the editor shall be held accountable for the use or misuse of the information in this book.

Introduction

Philip N. Wheeler

This is the fourth edition of *Alchemical Symbols*. The purpose of this revision is to add more symbols, correct a small number of errors, and improve the overall layout of the text. We believe this to be the most comprehensive presentation of alchemical symbols available.

Portions of the following symbol tables appeared in the "Last Will and Testament" of Basilius Valentinus in the seventeenth century. That table has been augmented with symbols from other sources including the "Alchemist's Handbook" by Frater Albertus, Dom Antoine Joseph Pernety's "The Great Art", Stanislas Klossowski de Rola's "Alchemy: The Secret Art," information taken from the works of Agrippa and John Read, symbols from the Bacstrom Manuscripts, and many other sources. For details please see the References section.

The second part of this work is a dictionary of basic Gematria. For those desirous of seeking possible hidden meaning in the alchemistical writings through the Qabalah, the Hebrew of key alchemically-oriented words is presented along with the numerical equivalences. Following that is a short Glossary of Latin Terms of interest to students of alchemy, tables of common conversion factors, and a list of abbreviations.

Alchemical symbols, or glyphs, are found in most of the writings on alchemy. Although many symbols can be quickly understood, an exact meaning for some symbols can be difficult to determine. Often it depends upon the context in which it is used. For example, the following symbol means Sun (Sol, Earth's sun) or Gold (Aurum; the metallic element Au):

This can make the study of symbol-laden alchemical texts more difficult. One should not rush to judgment on the meaning of a particular symbol in a given passage or as part of an emblem. The symbols must be read and interpreted within the context of the original work. This is especially important given the fact the text of many books were typeset by hand; I have seen examples where a symbol was inverted. For example, ⊻ instead of ⊼, which I have seen in original printings of Hazelrigg's "Book of Formulas."

I have arranged the symbols such that when there are several symbols with essentially the same meaning, the one that I believe to be the most common is shown first. Almost every symbol or symbol group starts with the symbols rendered by a vector-graphics computer program, and thereafter follows an example of the symbol(s) as painstakingly collected from Alchemical texts.

Finally, Hans Nintzel's "A French Alchemical Romance and Adventure" is included, its first appearance in print.

Philip N. Wheeler

Alchemical Symbols

Prima

☿	**Mercury**	Mind.
⊖	**Salt**	Base matter or body.
🜍	**Sulphur**	Spirit.

Classical Elements

🜁	**Air**	Hot and Wet. Sanguine.
🜃	**Earth**	Cold and Dry. Melancholy.
🜂	**Fire**	Hot and Dry. Choleric.
🜄	**Water**	Cold and Wet. Phlegmatic.

Planets and Metals

☉	**Sun**	Gold
☽	**Moon**	Silver
☿	**Mercury**	Mercury
♀	**Venus**	Copper
♂	**Mars**	Iron
♃	**Jupiter**	Tin
♄	**Saturn**	Lead

Zodiac

♈	**Aries**	Calcination.
♉	**Taurus**	Congelation.
♊	**Gemini**	Fixation.
♋	**Cancer**	Dissolution (Solution)
♌	**Leo**	Digestion.
♍	**Virgo**	Distillation.
♎	**Libra**	Sublimation.
♏	**Scorpio**	Separation.
♐	**Sagittarius**	Incineration.
♑	**Capricorn**	Fermentation.
♒	**Aquarius**	Multiplication.
♓	**Pisces**	Projection.

A

Acētum	Acetum is the Latin word for vinegar. Vinegar consists of from 3 to 9% acetic acid.

Acētum distillatum	Distilled vinegar.

Aes Ustum	A copper or brass cremator.

Air	Hot and Wet. Sanguine. One of the 4 classical elements. Latin: Aer.
Alcohol	Ethanol, C_2H_5OH, often distilled from wine. A volatile, flammable, colorless liquid. Sometimes referred to as Spirits.
Alembic	Alembicum Vitrum, a glass vessel for laboratory work.
Alkali, common	Can be Salt of Kali, Kali Mur, Potassium Chloride.

Alkali, common fixed	A fixed common alkali. See **Fixation**.	
Alkali, common volatile	Possibly ammonia.	
Aludel	A pot used for sublimation.	
Aluminum	Alumen. Chemical element with symbol Al and atomic number 13.		

Amalgam	Rebis; double matter. Mercury alloy.
Amalgamate	Amalgamare. Amalgamation.
Ammoniac	Sal ammoniac. Ammoniatum. Ammonium carbonate (Hartshorn), or Ammonium chloride.
Air	Hot and Wet. Sanguine. One of the 4 classical elements. The Latin word is Aer.

Ana	In equal parts.	
Annus	Annual. Year.	
Antimony	Element: Sb (Latin: stibium); atomic number: 51.		
Antimony, Oil of	An alchemical preparation.	
Aqua	Water. Chemical compound H_2O.		

Aqua Fortis	Nitric acid, HNO_3.	
Aquarius	Multiplication. Sometimes this might be salt nitre.	
Aqua Regia	A mixture of nitric acid and sulfuric acid.	
Aqua Vitae	Water of Life	

Arena	Sand.	
Argentum	Latin word for Silver. The Moon. Luna.	
Argentum Vivum	Quicksilver. Also ssee Mercury.		
Aries	A Zodiac sign. Calcination. Rarely means Antimony.		

Arsenic	Arsenicum. Chemical element with symbol As.		
Ashes	Cinis is Latin for Ash.		
Asphaltum	A form of asphalt (or bitumen) with a relatively high melting temperature. Also see Congelation.		

Attramentum	Black ink.		
Aunus	Unknown.		
Aurichalcum	Orichalcum. A type of bronze or brass metal alloy known as early as 400 B.C.		
Aurum	Latin word for Gold. Sol. The Sun.		
Aurum Pigmentum	Gold color.		

Aurum Potable	An alchemical preparation of gold.		
	⚲	⚳	
	⚴	⚵	⚶
Azoth	Comparable to the symbol for caduceus, the staff carried by Hermes in Greek mythology.		
	☿	☿	

B

Balneum Maria	BM. Balneum Mariæ. Water Bath. Bath of Mary.
	MB · MB
Balneum Vaporis	BV. Steam bath.
	VB · VB
Bismuth	Element with symbol Bi and atomic number 83.
	8 · 8
Black Ink	Also see Attramentum.
	[ ] · ⎕
Boil	The rapid vaporization of a liquid.
	(symbol)
Bone	A hard tissue; part of the vertebrate skeleton.
	tn · tn

Borax	Sodium borate, sodium tetraborate, or disodium tetraborate.
Brass	Metallic alloy of copper and zinc. Aurichalcum. Orichalcum.
Brick	Typically made of clay. Sometimes pulverized.
Bull	Astrological sign Taurus. Congelation.

C

Calcination	An alchemical process. The first symbol below is astrological Ares.

Calcine, to	An alchemical process.

Calx	Oxide.

Calx vive	Quicklime. Burnt lime. Calcium oxide (CaO).		
Camphor	A terpenoid with the chemical formula $C_{10}H_{16}O$	
Cancer	Dissolution: the process of dissolving a solute into a solvent to make a solution. An astrological sign.	

Capricorn	Fermentation: a metabolic process that consumes sugar in the absence of oxygen. An astrological sign.
Caput Mortuum	Dead head. Substance left over after a chemical operation such as calcination.
Carbonate	A salt of carbonic acid.
Cera	Cera alba; bee's wax.

Ceruse	Lead acetate.		
	‡	✚	T
	✚	✚	T
Chalk	A form of limestone composed of the mineral calcite.		
	C	C	
Ciment	French word that means Cement.		
	2	2	
Cinceres clavellati	Crude potassium carbonate		
	ᴪ	ᴪ	
Cinis	Ashes.		
	⊢E	☉	
	E	☉	

Cinnabar	Cinabris. A brick-red form of mercury sulfide.		

Coagulate	Process by which a substance changes from a liquid to a gel.		

Cohobate	The process of distillation of matter, with the liquid drawn from it; that liquid being poured again upon the matter left at the bottom of the vessel and the distillation repeated a number of times.	

Compose	Preparation.
Congelation	A process by which matter congeals.
Cool	To reduce a substance's temperature.
Copper	Cuprum. Venus.
Copper, Burnt	A preparation made from copper.

Cornua	Cervi. Hartshorn. Sal Ammoniac.	
		CC
		CC

Crocus Marti	An oxide of iron.		

Crocus Vereri	An oxide.		

Crucible	Crucibulus.		
Crystal	Crystalli.		
Curcurbite	The lower part of a distillation still.		

D

Decompose	Decay or cause to decay.
Dies	Latin word for Day.
Digerere	Separate.
Digest	Digestio.
Digestion	Digest in a moderate heat, such as in a B.M. (Leo)

Dissolution	Process associated with astrology sign Cancer.
Dissolve	A solid is mixed into a liquid, creating a solution.
Dissolvere	To loosen or destroy.
Distill	Distillare.
Distillation	Process associated with Virgo

Drachma	Dram. 1/8 ounce; 60 grains. See "Conversion Factors."	
	3	3
Drop		
	gt	gt
Dust	Pulvis.	
	O⧺	O⧺

E

Earth	Terra. Cold and Dry. Melancholy. One of the four classical elements.

Element	One of the primary constituents of matter.

Equal amounts	Indicates using substances in equal amounts.

Essence	Essential oil.

Evaporate	Evaporare.

Extract	Extrahere.

F

Fermentation	Process associated with Capricorn.
Ferrum	Iron. Associated with Mars.
Feu de Roue	Furnace.
Filings	Often filings of iron.

Filter	Filtrare.
Fimus Equinus	Digestion in horse dung or by a gentle fire such as B.M.
Fire	Ignis. Hot and Dry. Choleric. One of the 4 Classical Elements.
Firune	Meaning unknown.

Fix	A process by which a volatile substance is changed into a form, often solid, that is not affected by fire.
Fixation	Process associated with Gemini.
Fixed	Fixum.
Flegma	Phlegm.
Flores	Flower. Oxide of a metal.
Flucre	Fluidic; fluxing.

Furnace	Ignis circulator.

G	
Gemini	Astrological sign. Fixation.
Glance	An ore from Sankt Joachimsthal, containing less than 1% silver.[1]
Glass	Commonly made from silicate (sand).
Glass container	Laboratory vessel. Also see Alembic.

---

[1] "The Greater and Lesser Edifier" by Johann Grashoff. R.A.M.S. Library of Alchemy Volume 11.

Glass of Talc	Specula; hematite.	
Glassertz	A lead ore from St. Arinaberg that is rich in silver.[2]	
Goat	See Capricorn.	
Gold	Sol. Sun.		

[2] "The Greater and Lesser Edifier" by Johann Grashoff. R.A.M.S. Library of Alchemy Volume 11.

Gold Pigment	Aurum Pigmentum.
Grana	Grain. See the section on Conversion Factors.
Green of Copper	Copper Chelate. Virido Acris.
Gum	A sticky substance used for binding.
Gutta	Drop.

H

Half	Latin word is Semis. Half of any quantity.
	SS. SS.
Handful	Manipulus.
	Ms. Ms.
Hart's Horn	Sal ammoniac.
	(symbols)
Hematite	A mineral form of Iron oxide.
	(symbols)
Herb	Herbs are typically plants with savory or aromatic properties.
	℈B ℈B

Hora	Time or hour.

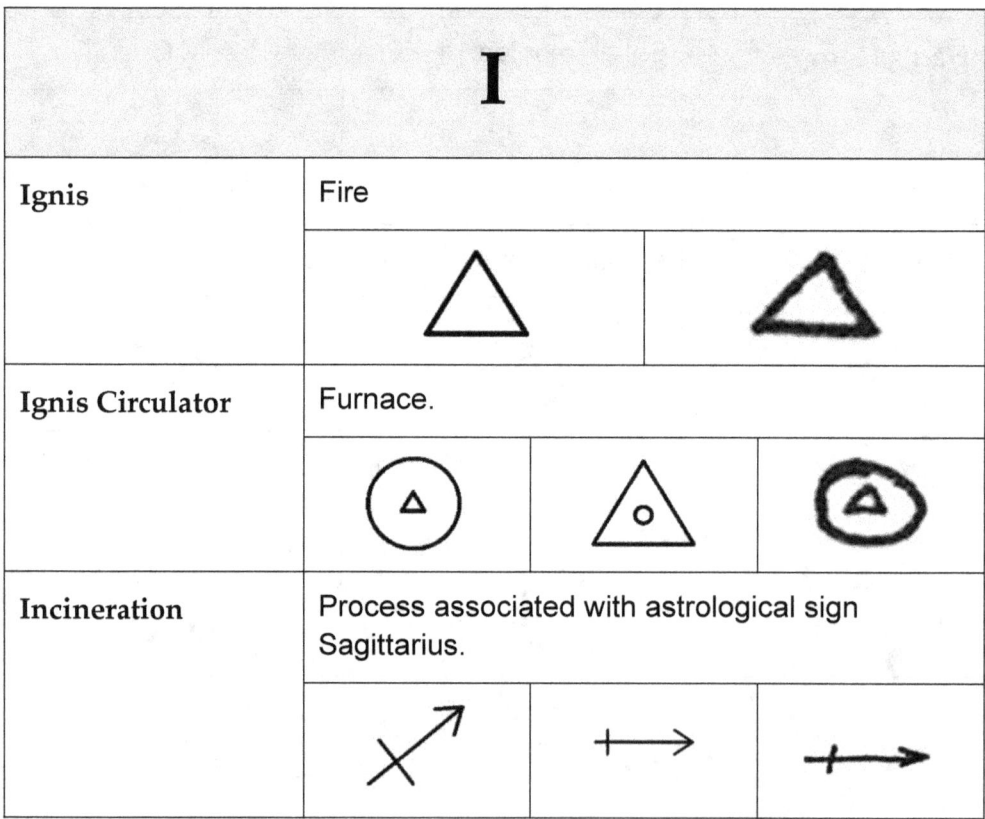

I

Ignis	Fire

Ignis Circulator	Furnace.

Incineration	Process associated with astrological sign Sagittarius.

Ink	A liquid or paste that contains pigments or dyes and is used to draw on a surface.	
Iron	Ferrum. Mars.	
Iron Filings	Small fragments of iron.	

J

Jupiter	Tin. Stannum.		

L

Lapis	Libra.

Latten	Laton.

Layered	Stratum super stratum. Layer upon layer.

Lead	Plumbum. Saturn.

Lead Acetate	Cerussa.	
Lead Monoxide	Lithargirium.	
Lead Oxide	Miny. Red Lead Oxide.	
Leo	Digestion.	
Libra	Scales, for weighing (pounds). Sublimation.	
Limatura Martis	Iron filings.	

Lime	Calx. Chalk.		
	C	G	ᶯ
	OᏦ	C	G
	ᶯ		OᏦ
Lime, Quick	Quicklime. Burnt lime. Calcium oxide.		
	♆	♯	↘
	⟋	⌐∘⌐∘	⚹
	⚹		↘
	⟋		∘⌐∘

Litharge	Lead monoxide, often colored red.		
Lixivium	Liquor.		
Lodestone	Magnet.		
Luna	Moon. Silver. Argentum.		

Lunaria	A lead ore, found near Freiberg in Meissen, that contains silver.[3]	
Lunarian	A lead ore, found at Villach, that is fairly pure lead.[4]	
Lute	Material used to seal vessels together during an alchemical operation. Hermetic seal.	
Lute of the Wise	See Lute.	

---

[3] "The Greater and Lesser Edifier" by Johann Grashoff. R.A.M.S. Library of Alchemy Volume 11.
[4] Ibid.

Luto	Lutrine. Mud of the Otter.	

M

Magnesia	Magnesium oxide.		
Magnet	Lodestone.		
Manipulus	Handful.		
Manure	Dung.		

Marcasite	Marcasito.		
Marriage	Join together.	
Mars	Iron.	
Martial Regulus of Antimony	Partially purified antimony.	

Materia	Latin for Material. Prima Materia is the First Matter.

Mercuriam Rubificati	A form of mercury.

Mercury	Argentum vivum. Quicksilver. Mercury is the integrative force, interweaving and balancing that of Salt and Sulphur. Circulation; dynamic equilibrium. One of the 3 Principles.

Mercury, precipitated	Mercury deposited in solid form from a solution.		
Mercury, sublimated	Sublimation is the transition of a substance directly from the solid to the gas phase, without passing through the intermediate liquid phase.		
Mercury of Saturn	Possibly a compound of lead and mercury.	
Miny	Red lead oxide.	

Month	Mensis.
Moon	Luna. Silver. Argentum.
Multiplication	Aquarius.

N	
Nickel	Chemical element with symbol Ni and atomic number 28.
Nitre	Saltpetre, Potassium Nitrate, Sodium Nitrate, soda niter.
Nitre, mined	Impure potassium nitrate.
Not Fixed	See entry for Fixed.
Nox	Night.

O

Oil	Oleum.

(symbols)	(symbols)	(symbols)

Oil of Vitriol	Made by distilling Green Vitriol (iron sulphate). This is sulfuric acid, a very strong mineral acid.

Orichalcum	Brass

Orpiment	A deep orange-yellow colored arsenic sulfide mineral.
Orpiment, Red	A mineral containing arsenic.
Oxide of a Metal	Flore.
Oxide, Zinc	A white compound that is insoluble in water.
Ounce	See the section on Conversion Factors.

Ounce, half	Uncia.

Ounce, eighth	Drachma. Dram.

Ounce, sixteenth	See the section on Conversion Factors.

P

Pisces	Projection.

Pitcher	Amphora.

Platinum	Chemical element, symbol Pt, atomic number 78.

Plumbum	Lead. Saturn.	
	♄	♄
	♃	#
Porous	Poras.	
	𐆐	𐆐
Potable Gold.	A gold solution that may be consumed.	
	♁	☦
	♁	☦
Potassium, crude.	Potash. Tartar, Cream of Tartar, Algol, Weinstein. Adjective of color is red or white depending on color of the wine that it came from.	
	♆	♆
Potassium Bitartrate	See Potassium, crude.	
	⊡	⊡

Potassium Carbonate	Sal tartari. Salt of Wormwood, called Pearlash if pure, makes 'Angel Water'. Salt of Tartar.	
Potassium Nitrate	Nitre, Nitrate of Potassa, Saltpeter, Stone Serpent.	
Potassium Sulphate	Salpo.	
Pound	See the section on Conversion Factors.		
Powder	Dust. Pulvis.		
Precipitate	Precipitare.		

Prima materia	First Matter.
	(prima)
Projection	Pisces.
Pulverize	Break a solid into tiny pieces.
Purify	Remove impurities.
Putrify	The breakdown of organic matter by rotting.

Q

Quantum satis	A sufficient quantity.		
	q s.	q ꝫ	q s.
Quantum vis	As much as you like.		
	q v.	q v.	
Quick Lime	Calcovviva.		
	♅	⯐	⌇
	⬦	⌁	⁎
	⌇	⬦	⌁
	⌇	♅	♅
Quicksilver	See Mercury.		
	☿	⊕	⊕
	☿v		☿

Quick Sulphur	A substance is said to be "quick" when it can neither die, nor be coagulated, nor congealed.[5]			
	⌂		⌂	
Quinta Essentia	Quintessence.			
	Q.E.		Q.E.	
	E	♀		Q
	E	♀		Q
	≠		≠	

R

Realgar	As arsenic sulfide mineral.		
Receiver	Vas Recipiens.		
Red Lead Oxide	Miny.		
Regulus	Pure metal.		

Alchemical Symbols

Retort	Retorta. Specialized laboratory glassware.		
	G	⌐	6
	G	ᴑ	6

Roots	Radices.	
	℟	℟

S

Sacharum	Sugar.	
	ff	ff

Saffrons of Mars	A calx of iron.	
	⊕−c	∋−
	⊕−c	∋−

68

Sagittarius	Incineration.

Sal Ammoniac	Also see Ammoniac.

Sal Gemma	Pure Salt. A salt mined in Poland.

Sal Prapuratum	Could be Saltpetre.

Sal Tartari	Probably an alkali.	
Salt, Alkali	Salt of Kali.	
Salt, common	Sal. The contractive force in Nature. Crystallization, condensation. One of the 3 Principles	
Salt, mined	Impure salt.	

Saltpetre	Sal Prapuratum.		
Salt, rock	Sodium chloride.		
Sand	Arena.		
Saturn	Lead. Plumbum.		

Scorpio	Separation.

Scruple	1/24 ounce. See the section on Conversion Factors.

Scruple, half	1/48 ounce.

Sifted Tiles	Flowers of Tiles. Also see Urine.

Silver	Moon. Luna. Argentum.

Silverglance	A lead ore found in Poland.
Silver-lead	A lead ore, from Hungaria, that contains silver.
Soda	Nitrum.

Sol	Sun. Gold. Compare with the symbols for Gold.

Sodium Biborate	Also see Borax.

Solid	Solidify.	
Solution	Solvere. To weaken, dissolve.	
Spirit	Sometimes refers to ethanol.	
Spirit of Wine	Ethanol.	

Spiritus	Vini Root.

Stannum	Tin. Jupiter.

Steel	Also see Mars and Iron.

Stone	Rock.

Stratum Super Stratum	Layer upon layer.

Sublimate	Sublimare.

Sublimation	Libra.		
Sufficient quantity	Quantum satis.		
Sugar	Sacharum.		
Sulphur	The Expansive force in Nature. Dissolution; Evaporation. One of the 3 Principles.		
Sulphur	Incorrect symbol for Sulphur.[6]		

[6] Seen in a printed copy of "The Golden Chain of Homer," and thought to be a typesetter's error whereby the Sulphur symbol was accidentally inverted. -pnw

Sulphur, Black	Possibly a dye made from sulphur.	
Sulphur, Quick	See Quick Sulphur.	
Sulphur, Sophic	One of the three essentials.	
Sun	Sol. Gold.		
Supo	Soap.	

T

Talcum	Talc.	
Tallow	A hard substance made from rendered animal fat.	

Tartar	Tartarus. Potassium Bitartrate.		

Tartar, Calx	Possibly an ash of tartar.	

Taurus	Congelation.		
Terra	Earth.		
Tigillum	Small beam.		
Tin	Stannum. Jupiter.		
Tincture	A solution, possibly a medicine.		

Tuna	A vessel for liquid.	
Tutty	Tutia.	
Twins	Gemini.		

U

Uncia	Ounce.
Universal Acid	
Universal Seed	
Urine	

V		
Vas Recipiens	Receiver.	
Venus	Copper.	
Ver-de-Gris	Green of Copper.	
Vinegar	Acetum.	

Vinegar, Distilled	Acetum distillatum.		
Virgo	Distillation.		
Vitriol	Vitriolum.		
Vitriol, Blue			

Vitriol, White	Zinc sulfate.	
Volatile	Not fixed. Active.	

W

Water	Aqua. Cold and Wet. Phlegmatic. One of the 4 Classical Elements.
Water Bath	Balneum Maria. BM.
Wax	Cera.
Work Completed	Finis.

Z	
Zinc	Chemical element with symbol Zn and atomic number 30.
Zinc Carbonate	Also could be Zinc Oxide. Tutia. Tuccia Preparatum.

A Table of Chymical & Philosophical Characters

by Basilius Valentinius

The table on the following page appeared in the "Last Will and Testament" of Basilius Valentinus (Basil Valentine), printed in London in 1657 and 1658.

It is generally believed that Basilius Valentinus was born in 1394 and that he joined the Benedictine Brotherhood, eventually becoming Canon of the Priory of St. Peter at Erfurt, near Strasburg, although we have no proof.

Basil Valentine was the originator of many chemical preparations of importance. Among these are the preparation of spirit of salt, or hydrochloric acid and oil of vitriol (sulphuric acid), and the extraction of copper from its pyrites by transforming it first into copper sulphate then plunging a bar of iron into the solution. He is also said to have discovered antimony.

His writings contain much of interest to the student of alchemy. Whether Basil Valentine is the correct name of the author or an alias does not matter since it detracts nothing from his work or the value of his experiments.

A Table of Chymicall & Philosophicall Charecters w.th their significations as they are usually found in Chymicall Authors both printed & in manuscript

Scheele's Chemical Symbols

by Carl Scheele

The four pages that follow show the tables of alchemical and chemical symbols used by Carl Wilhelm Scheele (1742 - 1786). These tables are from the work by Henrik Theophil Scheffer, *Chemiske föreläsningar,* Upsalla: M. Swederi. 1775.

Scheele, a German-Swedish alchemist and chemist, discovered oxygen before Joseph Priestly published his findings on oxygen. Scheele is said to be the first to discover other chemical elements, including Barium, Chlorine, Manganese, Molybdenum, and Tungsten. He also discovered various chemical compounds through his experiments, including citric acid, lactic acid, prussic acid, and hydrogen fluoride.

⊖ *Sal in genere*

✝ *acidum; c concentratum; d dilutum*

✝ *m. Acidum minerale*

✝🜨 *Acidum Vitrioli*

✝🜨 *c concentratum, d. dilutum*

✝🜕 *Acidum Nitri,* 🜕 *a n.phlogisticatum*

🜅 *Aqua fortis*

✝⊖ *Acidum Salis* ⊖ *a s dephlogisticatum*

🜆 *Aqua Regis*

✝🜟 *Acidum fluoris mineralis*

🜎 *Acidum Arsenici*

✝v. *Acidum Vegetabile*

✝🜊 *Acidum tartari*

✝🜨 *Acidum Sacchari*

🜋 *Acetum*

✝a. *Acidum animale*

✝▢ *Acidum urinæ; phosphori*

✝🝛 *Acidum formicarum*

🜹 *Acidum aereum; atmosphæricum*

⊕ *Sal alcalinus*

⊕p. *Sal alc. purus (Causticus)*

⊕v. *Alcali fixum vegetabile*

⊕m. *Alcali fixum minerale*

⊕ *Alcali volatile*

🜨 *Terra*

🜨 *Lapis*

⸫ *Arena*

🜨 *Calx, p. pura (ustulata)*

🜨 *Calx vitriolata (selenites, gipsum)*

🜨 *Terra ponderosa*

🜨 *Magnesia*

⊕✚ Sal neutralis

❶ Nitrum

⊖ c. Sal communis

♉ Tartarus: r. ruber, a. albus, p. purus

⬠ Borax

⊖⋈ Sal ammoniacus

♉✚ Sal medius terrestris cum acido

♉⊕ Magnesia vitriolata (Sal amarus anglic.)

○ Alumen

♉⊖ Sal medius terrestris cum alcali

♉⊕ Alcali volatile magnesia Saturatum

♛✚ Sal medius metallicus cum acido

♀♁ Vitriolum cupri (v. Coeruleum)

♁♂ Vitriolum ferri (v. viride)

♁♂ Vitriolum zinci (v. album)

☽❶ Luna nitrata (crystalli lunæ)

☿ Mercurius Sublimatus corrosivus

☿ Mercurius præcipitatus albus

♄✚ Plumbum acetatum (Saccharum Saturni)

⊕ Cuprum acetatum: ⊛ Cupr. acet. purum

☉♖ Aurum regalisatum

♛⊖ Sal medius metallicus cum alcali

♀⊕ Alcali volatile cupro saturatum

♀ Sal sedativus

△ Resina

♈ Gummi ✳ Gummi resina

♈ Metallum Sulphuratum

☿ Mercurius sulphuratus, (cinnabaris)

☉ Aurum (Sol)

◉ Platina

☽ Argentum (Luna)

▽ *Argilla*
♆ *Terra Silicea*
XX *Cryſtallus*
✖ *Vitrum*
⬭ *Fel vitri*
Λ *Minera*
♉ m. *Calx metallica*
⚶ *Aurum fulminans*
⚶ *Turpetum minerale*
⬤ *Arſenicum album*
♄✖ *Vitrum plumbi*
⊟ *Magnes*
♋ *Metallum*
Ɛ *Cinis*
▣ *Urina*
▽ *Aqua*
⛛ *Äer, n. Nudus*
△ *Ignis*
♧ *Phlogiſton*
♠ *Sulphur*
⚶ *Phosphorus*
⚶ *Pyrophorus*
✴ *Carbo*
☉ *Oleum unguinoſum*
⚭ *Oleum empyreumaticum*
⁖ *Oleum eſſentiale*
⁙ *Aether*
♈ *Spiritus vini* ℞ ſ *rectificatisſimus*
⊕♧ *Alcali phlogiſticatum*
⊕♠ *Hepar akalinum* ⚶ *Hepar terræpond*
⚶ *Hepar calcis;* ⚶ *Hepar magneſiæ*

☿ *Hydrargyrus (Mercurius)*

♄ *Plumbum (Saturnus)*

♀ *Cuprum (Venus)*

♂ *Ferrum (Mars)*

♃ *Stannum (Jupiter)*

♁ *Vismutum*

⚷ *Niccolum*

•–• *Arsenicum*

♨ *Cobaltum*

♋ *Zincum*

♆ *Antimonium*

♅ *Magnesium*

♏ *Retorta*

♍ *Recipiens*

♒ *Cucurbita;* ♒ *Alembicus*

♎ *Crucibulum*

⁑ *Evaporare* ☷ *tio. Evaporatio*

••• *re Digerere;* ••• *tio. Digestio*

≋ *re Coquere, ebullire*

⌒ *re. Destillare*

℈ *re. Incinerare*

♅ *re. Calcinare*

◗ *re. Sublimare;* ◗ *Sublimatum*

♌ *re Pulverisare,* ♌ *pulvis*

∿ *re. Solvere;* ∿ *tio Solutio*

◐ *re. Præcipitare;* ◐ *Præcipitatum*

♌ *re. Fundere;* ♌ *to fusio*

◉ *Residuum, Caput mortuum*

▥ *Mensis*

♂ *Dies;* ♀ *Nox;* ♂♀ *Nyctemeron*

♘ *Hora*

Table of Symbols from *Dissertation on Elective Attractions*

By Torbern Bergman, 1775

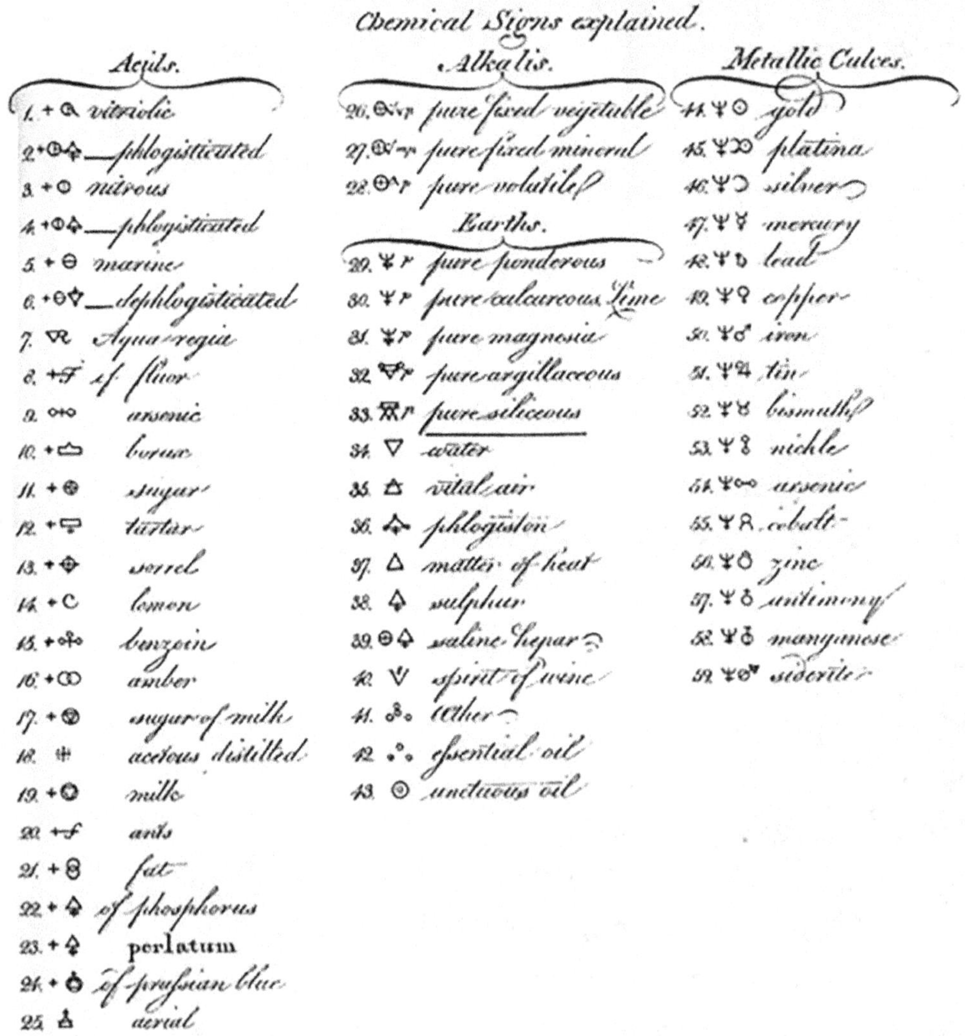

Chemical Signs explained.

Acids.

1. + ♈ vitriolic
2. +⊙♉ ___ phlogisticated
3. + ⊙ nitrous
4. +⊙♉ ___ phlogisticated
5. + ⊖ marine
6. +⊙♉ ___ dephlogisticated
7. ℞ Aqua regia
8. +♒ of fluor
9. ⚯ arsenic
10. + ⌂ borax
11. + ⊕ sugar
12. +♆ tartar
13. + ⊕ sorrel
14. + ☾ lemon
15. +♣ benzoin
16. +∞ amber
17. + ❀ sugar of milk
18. ♯ acetous distilled
19. +⊙ milk
20. +♐ ants
21. + 8 fat
22. + ♃ of phosphorus
23. + ♃ perlatum
24. + ♠ of prussian blue
25. ♄ aerial

Alkalis.

26. ⊙♐ pure fixed vegetable
27. ⊙♐ pure fixed mineral
28. ⊙♐ pure volatile

Earths.

29. ♉♈ pure ponderous
30. ♉♈ pure calcareous Lime
31. ♉♈ pure magnesia
32. ♉♈ pure argillaceous
33. ℞♈ pure siliceous
34. ▽ water
35. △ vital air
36. ♣ phlogiston
37. △ matter of heat
38. ♣ sulphur
39. ⊙♣ saline hepar
40. ♈ spirit of wine
41. ♂ æther
42. ∴ essential oil
43. ⊙ unctuous oil

Metallic Calces.

44. ♀⊙ gold
45. ♀∞ platina
46. ♀☽ silver
47. ♀♉ mercury
48. ♀♄ lead
49. ♀♀ copper
50. ♀♂ iron
51. ♀♃ tin
52. ♀♉ bismuth
53. ♀8 nickle
54. ♀∞ arsenic
55. ♀℞ cobalt
56. ♀♄ zinc
57. ♀♂ antimony
58. ♀♄ manganese
59. ♀♂ siderite

Gematria

The purpose of the following dictionary of gematria is to provide a convenient reference for those interested in pursuing the possibility of numerical codes in the Alchemistical writings. These entries are no means complete and are the Hebrew words only pertaining to alchemy. The possibility of Greek and Latin gematria certainly exists but the reader must pursue his own devices in this regard. For the Hebrew alchemical terms, the English word is given in the first column, the anglicized Hebrew in the second column and the numerical equivalent in the third column. Following the third column, a Biblical reference might be given or some other note that may be useful. In some cases, there are several words in Hebrew that convey the same sort of meaning. In most cases of this sort, the various choices are all given. The user will find Sepher Sephiroth and a Hebrew dictionary most useful here. It is hoped that the below offering will prove useful to those fond of these pursuits.

The first table below, Basic Gematria, gives the decimal equivalent for each letter of the Hebrew alphabet.

The second table below gives the Anglicized equivalent for a particular Hebrew letter as used in the gematria dictionary. The numerical value is also given. In the anglicized word, no attempt was made to delineate various final letters having particular numerical value. Instead, the "normal" form of the Hebrew letter has been used.

Basic Gematria

Decimal	Hebrew	Glyph
1	Aleph (Alef)	א
2	Beth (Bet)	ב
3	Gimel	ג
4	Daleth (Dalet)	ד
5	He (Hei)	ה
6	Vau (Waw)	ו
7	Zayin	ז
8	Heth (Cheit)	ח
9	Teth (Teit)	ט

10	Yod (Yodh)	י
20	Kaph (Kaf)	כ, ך
30	Lamedh	ל
40	Mem	מ, ם
50	Nun	נ, ן
60	Samekh	ס
70	Ayin	ע
80	Pe (Pei)	פ, ף
90	Tsadi (Sadhe)	צ, ץ
100	Qoph (Qof)	ק
200	Resh (Reish)	ר

300	Shin	שׁ
400	Taw (Tav)	ת
500	Kaph (Kaf)	ך
600	Mem	ם
700	Nun	ן
800	Pe (Pei)	ף
900	Tsadi (Sadhe)	ץ

Anglicized Hebrew Letters

כ	י	ט	ח	ז	ו	ה	ד	ג	ב	א
K	Y	T	Ch	Z	V	H	D	G	B	A

ת	שׁ	ר	ק	צ	פ	ע	ס	נ	מ	ל
Tv	Sh	R	Q	Tz	P	O	S	N	M	L

Gematria Dictionary

Abraham	ABRHM	248
Acid	ChMTzH	143
Acid, Sulfuric	ChMTzH GPRN	476
Adam	ADM	45
Adam Kadmon	ADM QDMVN	245
Age (duration)	OVLM	146
Agla	AGLA	35
Air	RVCh	214
Air, the	AVYR	217
Alchemy	ALKMH	106
Alcohol	MSQH	205
Alkahest	MTvYK HKL	525
Alkali	QLY	140
	AShLG	334
Amalgam	TvORBT KSP ChY	1250
	HRKBD	231
	AMLGMH	119
Amalgamate	ORB	272
	HRKB	227
	HTvAChD	418

Amber	ChShML	378
Angel	MLKA	91
Anoint, to	MShCh	348
Antimony	PVK	106
	TzDYD	108
	KChL	58
Aquarius	DLY	44
Arcana	RZIA	218
Arise, to	QVM	146
Ashes	APR	281
Ate	AKL	51
Auriel	AVRIAL	248
Azoth	AZVTh	414
Chesed	ChSD	72
Chokmah	ChKMH	73
Circle	ZGVL	109
Clean	THVR	220
	THR	214
Closed	AThM	50
Cloud	BMH	47
Coagula	QBA	103
Coagulate	QPA	181
	QRSh	600

	HQPA	186
	HQRSh	605
Coctio	BYShL	342
Cohesio	DBQVTh	512
Cohesion	ORVTv	676
Colors	GVVNY	75
Conceive, to	HRH	210
Concentrate	RKZ	227
Consume	ChML	98
Consuming	AVKL	57
Copper (brass)	NChShT	758
Corpse	GVPH	94
	NBLH	87
Corrosive	MAKL	91
	MBLH	77
	MPSYD	194
Corrupted	MKH	65
Create	BRA	203
Created	BAR	203
Crucible	MTzRP	410
	KVR	226
	TzLYL	160
Crystal	GBYSh	315

Cup	KVM	86
Darkness	AVPL	117
Death	MVT	55
	MT	49
Destroy, to	MKH	65
Destruction	ThL	400
Dew	TL	39
Dissolve	MSH	105
	ShYM QTZ LY	506
	HTvR	605
	HMChH	58
Dissolution	BYThYL	57
Distill	HTP	94
	TzRP	370
	QPR	380
	ZLP	117
	ZQP	187
Distillation	ZQVQ	213
	TzRVP	376
Dragon, the	ThN	450
Drink, to	ShTvH	705

Drop, a	ThPH	94
Drops	AGLY	44
Dulcify	HNYM	105
Dung	GLL	63
Dust	OPR	350
Eagle (Aquila)	QZNYH NShR	722
	NShR	550
Earthenware Vessel	KD	20
East	QDM	144
Egg, an	BYZH	107
Elevated or Exalted	GBH	10
Elevatus	NShR	550
Enchantment	LHTY	54
Essence, the	ATh	401
Ether	AVYRA	218
Evil	ROH	275
Experiment	DChVN	66
Extinguish, to	DOK	94
Fæces (globular)	GLL	63
Fecundity	AChLB	41
Ferment, to	HMH	50
	ChMTz	118
Fiery	AShY	311

	DVOR	280
	LVHP	121
Fiery Serpent	ShRP	580
	NChSh	358
Fire	ASh	301
Firebrand	AVD	10
Fish	DG	7
Flame	LHB	37
	LHT	44
Flask	DMNVN	150
	PK	85
	DQDVQ	214
Flew	RAH	10
Flood, a	YAR	211
Flower, a	GBYO	85
Flux	ZRM	247
	ShTP	389
	ZYRMH	262
Fly	BRCh	210
	OVP	156
	HADR	210
Fountain	MOYYN	180
Furnace	ThNVR	656

	KBRVH	233
	KBShN	372
	KVR	226
Gathered	AMP	141
Geburah	GBVRH	216
Gemini	ThAVMYM	497
Giving up	MVLCh	104
Glass	ZKVKYT	72
Glory	GAVN	60
Gluten	DBQ	106
Gold, shining	ZHB	14
Gold, pure, fine	KTM	69
Gold, with Silver	BTzR	292
Gold, dug up	ChRVTz	304
Gold, pure	ZHB ShChVT	337
Gold, shut up	ZHB SGVR	283
Gold, red	ZHB PRVYM	344
Gold, good	ZHB TVB	31
Gold, of Ophir	ZHB AVPYR	311
Gold, water of	MY ZHB	64
Great	GDVL	43
Great Dragon, the	ThLY	440
Great Stone, the	GDLH ABN	95

	YRVQ	316
Green	LCh	38
	RTzNN	370
	DSh	304
Green color (herbaceous)	IRQ	310
Guard	GNN	103
Harden, to	KBD	25
Hardness	ChVSN	124
Healing	AMA	62
Heart	ChVL	44
Heat	ChM	48
Heaven	ShMYM	390
Herbage	TzShB	372
Hod	HVD	15
Holy	QVRSh	410
Honey	DBSh	306
	BShH	306
Humour	RTzNN	390
	MLA	71
	BQShT	411
Ice	QRCh	308
Impure	ThMA	50

increase, an	MVSP	186
Internal	PNYMY	190
Iron	BRZL	239
	BRTzL	322
Jars	TzNThRVTh	1146
Jeruschalim	YRVShLYM	596
Jesus	YHShVH	326
Jupiter	TzDQ	194
Kether	KTR	620
Kill, to	HRG	208
King	MLK	90
Knowledge	YDYZH	99
Lead	OPRT	750
Lead (a plummet)	ANK	71
Leopard	NMRA	291
Life	CHYYM	628
	NShMH	395
Lifted up	ZQP	187
Light	AVR	628
Lightening	BQZ	109
Limpid Blood	TzVL	156
Lion	ARYH	216
Lion, fierce	LYSh	340

Lion (a whelp)	GVR	209
Lion, young	KPYR	310
Lioness	LBYA	43
Lover, a	MAHBH	53
Lucifer	HYLL	75
Lute (obstruct, shut up)	ChMN	108
Magician	ChRTM	257
Male	DKVRA	237
	ZKR	227
Malkuth	MLKVT	496
Marriage	ChTvNH	463
	NShVATM	407
Mars	MADYM	655
Matter	GShM DBR	549
Material	MVTKTz	127
Measure	MDH	49
Measure out	ChQ	108
Meditated	DMK	49
Melt	NZL	87
Mercury (the planet)	KVKB	48
Mecury (quicksilver)	KSP ChY	178
Metal	OZM	117

	TvMZYTv	857
	MTvKTv	860
Metallic Oxide	ChMTzTv MTvKTv	1398
Metallic Salt	MLCh MTvKTv	938
Milk	ChLB	40
Moon (full)	LBNH	87
Moon, the	YRCh	218
Mother	AM	41
Mouth, the	PH	85
Mud	BZ	92
Myrrh	MR	240
	MVR	246
Nechesh	NChSh	358
Netzach	NTzCh	148
Night	LIL	70
Nitre (saltpeter)	NTR	259
Obstruct (shut up)	ChMM	108
Odor	RYCh	218
Oil	YZHR	222
Olive	ZHTTv	421
Passive	MQBYL	182
Penetrate (to be sharp)	ChD	12

Perished	ABR	203
Phallus	ShPKH	203
Phlegm	BLGM	55
	KYCh	38
	LChTv	128
Philosophical	PYLNSNPY	370
Philosopher's Stone	ABN HPYLNSNPYM	468
Planet	KVKB	48
Pool, a	AGM	44
Pour out, to	YTzQ	200
Power	KCH	28
	ChYLY	58
Pure	BR	202
	YShR	510
	TZCh	24
	ZK	27
	RQB	302
Purified or Purification	BRR	402
Purple	ARGMN	294
Putrifaction	BASh	303
	RQBNNM	442

	BLY	42
	RQB	302
Putrify	HRQB	307
	HBASh	308
Queen	MLKH	95
Quicklime	SYD HY	83
	GYR	213
Quicksilver	KSP ChY	98
	KTPChY	127
Quintessence	ChMSh TVShYH	678
Ram, a	ThLH	44
Receive, to	QBL	132
	TvQN	550
Rectify	TZRP	296
	ZQQ	207
	ADM	45
Red	SMQ	200
	ShRQ	600
Regulus	DN MLK	144
Remove, to	ND	54
Rest	NYChCh	76
Retort	TvShBH	713
	KHLKH	80

Rock	ABN	53
Salt	MLCh	78
Salt, Sea	YM MLCh	168
Sand	ChVL	44
Saturn	ShBTAY	78
Sea, the	YM	50
Secret	NSTzRH	715
Seed, a	QNH	155
Seethe	ZVD	17
Seething	NChL	88
Semen	ZRO	277
Separation	HBDLH	46
	PYRVD	300
Serpent	NChShT	357
Sharpness	ChDQ	112
Silent	DMH	49
Skull, a	GLGTv	466
	GLGLTv	473
Silver	KSP	160
Smooth	PShVTv	786
Snow	ShLG	333
Soar	RChP	288

Son, the	BN	52
Soul	NPSh	430
Spikenard	NRD	253
Spirit	RVCh	214
Star	KVKB	48
Star, morning	HYLL BN ShChR	635
Star, day	KVKB HBQR	355
Steam (vapor)	AID	15
Strength	OZ	77
Suction	YNYQH	175
Sulfur	QPYRTv	693
Sun, the (Sol)	ShMSh	713
Supernal Mother	AYMA	52
Sweet	MTvVQ	546
Swell, to	HYM	55
Tile	LBYNH	97
Tin	BDYL	46
	GVN	59
Tincture	TOM	117
	AYKVTv	437
Tinge	NTvN	500
	TzBO	162

Tipareth	TvPARTv	1081
Trial, a	BChVN	66
Union	ODH	79
	YChVR	224
Universal	KLL	80
Urine	ShTN	359
	SYN	120
	HShTvNH	760
	MY RGLYM	333
Vacuum	RYQM	350
Vapor (steam)	AID	15
Vases or Vessels	KLYM	100
Vehement	ChZQ	115
Venus	NVGH	64
Vinegar	ChMO	118
Viper, a	APTzH	156
Virgin, a	MLKA	91
Vitriol	QLQNTvNM	770
	ZG	10
	ZTRYNL	306
Vitriolic	ZTRYNLY	316
Vitrify	HPK LZKNKYTv	642
Volatile	QL RASh	631

	HPKPK	205
Watchtower	BChN	60
Water	MYM	90
Water Pot, a	KD	20
War	MLChMH	123
Whelp (of a lion)	GVR	209
White	LBN	82
	ZCh	15
Whiteness	LBNH	87
Wife (woman)	NYQBH	163
	AShH	306
Wine	YYN	70
Wise (wisdom)	ChKM	68
Wolf	MOShH	415
	MLAKH	96
Yeast	ShANR	507
Yesod	YSVD	80

Glossary of Latin Terms

Aqua fortis	Strong water
Aqua Mercurialis	Water of Mercury.
Aurum potabile.	Drinkable gold
Aqua regia	Royal water
B.M.	Balneum Mariæ. Mary's bath (i.e. water bath).
Balneum	Bath
Butter of antimony	Antimony trichloride
Caput Cervi	Crow's head
Conjunctio Spiritus cum Corpore	The marriage of the spirit with the body
Duplex	Two-fold; double
Duplicatus	Doubled
e.g.	For example
Elixir Rubrum Naturae	The indeterminated red elixir

indeterminatum	of nature
Elixeria tertia ordinis	(The) Elixirs of the third order
Ex Mercurio	From Mercury
Exuberatus	Proliferated.
Fermentum Lunae	Lunar ferment
Finis	The end
First in mercurio quidquid quaerunt Sapientes	Whatever the Knowing Ones seek is in (the) mercury.
Flowers of antimony	Antimony trioxide Sb_2O_3
Forma	The appearance.
Fusibilis	Fusible.
Gluten Aquilae	Gluten of the eagle.
i.e.	In other words; that is
Ignis Gehennae.	(The) Fire of Gehenna (Hell)
In balneo	In the bath.
In forma Aquaé, olei or	In the form of water, oil or

butyri	butter.
In forma metallica currente	In the form of a running metal.
In forma olei	In the form of an oil.
Indugator.	One who brings forth water; one who directs the flow of water
Lac Virginis.	Virgin's milk.
Lapis Albus	White Stone.
Lapis calaminaris	Zinc silicate (hemimorphite)
Lapis Sophorum medicinalis universalis	The Wise Ones' stone of universal medicine.
Lixivium	A solution of alkaline salts
Luna cornea	The horned moon
Mercurius Sophorum animatus	The spirited Mercury of the Wise.
Natura Media.	Natural means
Oleum Lunae	Oil of the moon

Oleum Solis	Oil of the sun
Oleum Suiphuris Lunae	Oil of the Lunar Sulphur
Per alembicum	By alembic
Per deliquium	By removal (of sediment); by clarification. (running into a liquid); through melting
Per ignem suppressionis	By the fire of concealment.
Per minima	At least; by the least
Per Se	By itself
Prjmae, secundae, tertjae ordinis	Of the first, second, third order
Quantum satis	Sufficient quantity
Quantum vis	As much as you like
Quinta essentia medicinalis	The fifth essence of medicine
Sal armoniac	Ammonium chloride
Soli Deo Gloria!	To God alone the glory!

Spiritus vini	Spirit of wine
Stibnite	Antimony trisulfide
Sulphur naturae album	The white sulphur of nature.
Sulphur Rubrum Naturae indeterminatum	The indeterminated red sulphur of Nature
Via humida	The humid way; by the humid way
Vide	See
Vitriol	A salt derived from sulfuric acid
Vitriol, blue	Copper sulfate
Vitriol, green	Iron sulfate
Vitriol, white	Zinc sulfate
Vivus	Living

Alchemical Symbols

Conversion Factors

Ounce variant	Grams	Grains
International avoirdupois ounce	28.3495231	437.5
International troy ounce = Apothecaries' ounce	31.1034768	480
Maria Theresa ounce	28.0668	
Spanish ounce	28.75	
Dutch metric ounce	100	
Chinese metric ounce	50	

ENGLISH APOTHECARIES' SYSTEM OF WEIGHTS				
Pound lb	Ounce ℥	Dram ℨ	Scruple ℈	Grain ℈℞
1	12	96	288	5,760
	1	8	24	480
		1	3	60
			1	20
373 grams	31.1 grams	3.89 grams	1.296 grams	64.8 milligrams

LIQUID		
1 fluid dram	1/8 fluid ounce	3.6969 mL (USA)
		3.5516 mL (British)
1 teaspoon	1/6 fluid ounce (USA)	4.93 mL (USA)
		5.0 mL (British)
1 tablespoon	3 teaspoons	

Abbreviations

aa   equal parts of each

B.M.   Balneum Mariæ; cook in a double boiler

e.g.   for example

ff   a reference. Example: "Pg. 300 ff" means the referenced area starts at page 300 and continues forward.

i.e.   in other words; that is

iij.   3 fluid drams

ij.   2 fluid drams

lbij.   2 pounds

lbj.   1 pound

Lot   1 lot = 0.5 ounce

p.   page

P.ij.   2 parts

P.iv.   4 parts

qs   quantum satis; sufficient quantity

qv   quantum vis; as much as you like

Rc.   Recipe

Rx.   Medical prescription

ss   half of any quantity

sv   spirit of wine

References of Interest

Albertus, Frater. *The Alchemists Handbook: Manual for Practical Laboratory Alchemy*. Red Wheel/Weiser, York Beach ME, 1974

Bacstrom, Sigismond. Bacstrom's Notebooks, Part 1 and Part 2. The R.A.M.S. Library of Alchemy, Vol. 7 and 8.

de Rola, Stanislas Klossowski. *Alchemy: The Secret Art*. Thames & Hudson, 1986

Gettings, Fred. *Dictionary of Occult, Hermetic and Alchemical Sigils*. Routledge & Kegan Paul, 1981

Hazelrigg, John. *The Book of Formulas*. Alembic Publishing, Waynesboro VA, 2011.

Nintzel, Hans. *Alchemical Symbols and Gematria*. R.A.M.S. Circa 1980

Pernety, Dom Antoine-Joseph. *A Treatise on the Great Art*. AGNZ, 1898

Unicode Standard   http://www.unicode.org/charts/PDF/U1F700.pdf

Valentinus, Basilius. *Last Will and Testament*. London, 1657

A French Alchemical Romance and Adventure

By Hans W. Nintzel

Richardson Texas, 1986

Since I promised many of my friends that I would recount to them the circumstances of my trip to France and sojourn with the Filiation Solazaref, I present this chronicle of how it came to be, the salient events of the trip, and its portents for the future.

It all started with a letter from a correspondent, Sig. Jose Anes, in Lisbon Portugal. In the letter he mentioned the Filiation Solazaref, and that they had published an outstanding book on alchemy, in French to be sure, and that I should look into it. I needed no further encouragement! Having secured their address, I wrote the Filiation and inquired about the book by one "Solazaref." Simultaneously, I wrote to a few other organizations in France concerning their publications. I didn't realize until later that these groups were publishing the Solazaref material as magazine articles. In particular were the "Verité Inderdite" and "Temepeté Chymique." Both had material written by Solazaref but not contained in his book.

I received a letter from the Filiation with an inquiry as to how I came to learn of the group, the book, and the Master (for such they call him) Solazaref. The letter stated that they were a private group, and that the book was privately published and not generally available to the public. If I desired more information, I should write to a certain Mme. Roux in Clermont-Ferrand, making a formal request for this data. This I did and first received a letter, in French, that described in broad terms the purpose of the book and some indication of its content and the author,

127

Solazaref. Immediately thereafter, I received a telephone call from Montreal from a Camille Coudari who introduced himself as a member of the Filiation "...and would I have time to talk?" WOULD I??? **SURE!!**

He asked me several questions about if I felt that more alchemical knowledge was being made available and if I had noticed that much of this knowledge was now coming from France. Affirmative! He confirmed a suspicion I had that the Soviets had taken, or have tried to take, Afghanistan because of the alchemical center in Nuristan. This, of course, made me warm up to the man!

He talked about the book and its contents at length. I was impressed. He asked about my work in the lab, my work with R.A.M.S. and so forth. He asked if I was interested in the book. My comments about the bear's personal habits in the woods died on my lips due to great modesty! I simply grunted: "Shoot, yeah"! He thereon asked if I would write him a letter outlining my alchemical activities and requesting the right to procure a copy of the book and the right to get it translated as I read almost no French. I agreed and he promised to hand-carry the letter to France and present it to the Master. He also suggested translating "L'Alchimie Expliquce Sur Ses Textes Classiques" & "Le Laboratoire Alchimque" by Eugene Canseliet and Atoréne, respectively. He told me both these books were **VERY** revealing and worthy of the "R.A.M.S. treatment". He agreed to purchase the two books on my behalf and send them to me. He did and I have them in my library now awaiting further action.

At this time, I was under a lot of stress. My business was not going very well, due to the oil crisis and I was really hustling trying to keep things going. In late July of 1986, I got a letter from France, from the

Filiation. It said that due to my communications with Camille Coudari, the Filiation would like, no, the MASTER, would like to meet me on the native soil of the Filiation. Once I arrived, I would understand why the trip across the Atlantic was necessary. So, if I could pick a day between August 3 and 14, they would arrange for a rendezvous!

Several questions popped to my mind. First, who would pay for the trip (although I suspected I already knew this answer!)? How long would I be there? What would the purpose of the trip be? And did the fact that I spoke virtually no French mean anything? Whew. If I was going to haul off for Clermont-Ferrand (wherever **THAT** was), I needed a little reassuring.

So, I wrote a letter in haste. I also had several personal problems, like how could I get off from work for this escapade. How could I possibly justify spending all that money when things were, well, "cash—flow—slim"? And so on.

While awaiting a reply from France, I discussed this with a good friend who has a genuine interest in the occult if not alchemy. My friend said: "I think you should go. There could be many implications." In my own sweet manner, I asked if he was going to buy me a round-trip ticket to France. He replied: "Well, if you don't mind going tourist class." I admitted I wouldn't mind! Seems he had accrued a lot of mileage under his American Airlines AAdvantage plan and was willing to let me cash one in for a round-trip. Whew. And he "booted in" a discount ticket for the Sheraton Hotel and a week of free car rental. Wow. I graciously accepted.

Other problems were solved by the Universe, which seemed BENT on my getting there! My job terminated. An inquiry for a speaking engagement was augmented by a sale of a large order of R.A.M.S. materials. Voila! (as they say) I now had the time and some spending money as well. With no further ado, I started making arrangements: Getting the ticket, calling Mme. Roux, etc. Everything, as it turned out, got down to "squeaking in" and "in the nick of time" situations.

The tickets required a certain amount of time to process. Not enough time, through the "regular" channels, as it turned out, for me to make it. So, with the help of some friends, we got some "inside" data and managed to go pick up the tickets (an 'impossibility' I was told!) the day before I would have to leave if I was to be able to spend at least five days or so with the Filiation. We did it. Also, a call to Madame Roux revealed I was quite lucky in catching her as she was about to join the Filiation in their "camp-in-the-woods" where there was no phone! Wow!!!

Now Mme. Roux speaks fair English except when she gets excited, which is about one minute into any series of communications! I asked her to "hang loose" and I called Nathalie in Colorado, a French translator. I explained to Nathalie what I wanted and had her call Mme. Roux with my message as to what I wanted done.

Specifically, this was to have Camille call me (he is fluent in English and French) so we could iron out details. The Thursday night, prior to my leaving the next Tuesday, he called at four AM, and the conversation was something like this: Camille: "Mr. Nintzel"? Yes. "When can you arrive in France?" On Wednesday morning at 9:45 August 13. "O.K. From Orly, take a cab to Paris to Gare d'Lyon. Take a train to Clermont-Ferrand where you will change to a train that takes you to Brioude. I will meet you at the train station in Brioude at 5:45 in the afternoon. Oh, do you have a sleeping bag? I ask because there may not be room in the house for you?" Well, uh, yes. "Good. Alright. See you next Wednesday." CLICK. Hello? Camille, Hello, operator, operator...hello. Silence reigned supreme and the die was cast!

First thing was the plane had a "little tinkering to be done on the electrical system." This caused an hour and a half delay. Realize, there is no phone where I can call Camille and Mme. Roux is now in the camp and not at home. And, before I left, a friend from India, Pierre, who had taken up summer residence in Lyons, wrote and said, "Don't bother coming. It will be a waste of time."

We made up ten minutes time and arrived an hour late. 10:30 am and the train to C-F was to leave at 12:07pm. First cabby wanted 400 Francs to go to Gare d'Lyon. "Trop Cher" said I. (too much). I got a guy who would go for 100 francs the going rate. Hit it son! We arrived and I was 'schlepping' two suitcases (one with my sleeping bag in it) and a "U-Tot'M" hanging around my neck. I mean we are talking **WEIGHT** here!!

At the train station I huffed and puffed to the information desk where, it turned out, they spoke 'Anglais'. In the queue rotation, I came up and panted, "I need a ticket for Clermont-Ferrand. Where..." "I" he said. "I"? Oui, "I"! and he pointed out to the great spaces. I espied "guichets" or ticket counters. They had letters A... M lit up on them. I ran to "I" and...no one home! I ran back to the info counter and (im)patiently awaited my turn. It was now around 11:30 am. I said: "There is no one at "I." He said: "Ten minutes." Ah. I 'hung out' for ten minutes. No signs of life and it was getting to be quarter of twelve. I ran back. "For God's sake, where else can I get a billet for Clermont-Ferrand"? "Any ticket window," quoth he. "But" I whined "You told me "I." "Train **LEAVES** on Track "I". You can buy a ticket at **any** window." Oy.

I huffed to a window and the clock spun inexorably on. If I miss the train, no way I can call and explain I am on a later train, right? RIGHT! This is the BIG one. I explain to the chap in front of me using my newly bought French-English-French dictionairé. He was sympathetic but not enough to let me go ahead. In fact, he wrote a check for his ticket and spent some time fetching suitable i.d. God! I finally bellied up to the Guichet and in my best French croaked: "Une billét pour Clermont-Ferrand, premier classe, sil vous plait." Ticket in hand I sped for track I. My nervous eye sought a clock. It was noon, straight up! Seven minutes to go before the choo choo went bye-bye without me!!!

Track I, was of course, at the far end. With my suitcases causing me to weave like a drunken Godzilla and my travel bag swinging rakishly from side-to-side, I hoofed it to track I. One train so no decisions to make. I jumped on...it started off!

I should mention, that before I left, I made a reservation for Tuesday night at the Sheraton-Montparnasse (at $160/night) in Paris. I also wrote to friends in Switzerland, Belgium, France, Paris, Pierre in Lyons (who I wished very much to meet), Andorra, Italy, Spain, Germany, Norway and Portugal. I invited all to meet me in Paris at the Sheraton and we would get together, have dinner and generally groove on one another's presence.

As the train slicked its way out of Gare d'Lyon, I slumped in a chair trying to regain my breath and slow my beating heart. I started to wonder how I could be so stupid as to do this. I could be home slipping into a nice dry martini and here I am huffing and puffing my way to some unknown destination.

The conductor's harrumph startled me and I handed over the ticket. He said something in French which I couldn't get. "Je Americaine. Je ne comprend pas," I said. "First Class," said he. "This IS a first class ticket," I huffed. "Teecket EES first class. Thees coach is second class." Oh. I trudged on into the night (at least to the next car).

I looked at the ticket. (Computer generated). It said: "Clermont-Ferrand via Nevers". Via Nevers? Did that mean I had to change at Nevers? Oh shoot. Well, I found a woman who knew enough English to make her dangerous and between her and my trusty dictionary we concluded this train went straight through to Brioude. In fact, she was "going out of Brioude" so just stick along. Alright! At Clermont-Ferrand we realized that, in my haste, I had not specified BRIOUDE. Uh oh. So, at C-F I had to schlep my bags to a guichet to get a ticket. Incidentally, I had been told, quite incorrectly, that everything in France was cheap due to the superior worth of the dollar. Ha ha. Accordingly, I went to Deake-Perrera and purchased only a thousand Francs worth of French currency. A mistake. By the time I bought the Brioude leg of the ticket, I had spent five hundred of my 1000 francs!! And a week to go! I felt queasy again.

Then, I couldn't find my way back to the lady who was "sponsoring" me. Too many passageways. I ran back and forth and finally found her. She motioned me to get on the train. It lurched forward. As I swung suitcases aboard and jumped, I noticed the terminal said that this train was going to...NIMES! I nearly had a stroke. I hurriedly sought my dictionary and inquired of the lady. Oh yes, this train WAS going to Nimes. Where she was, in fact, going. But, but, I thought this was the train to Brioude. Oh yes, it goes to Brioude. But, which is it? Nimes or Brioude? Both, she said firmly. I suddenly realized, it must stop at Brioude on the way to Nimes! (as was the case.)

I sat down and the lady next to me started a conversation. "Je ne comprend." She shrugged and stared out the window. The lady (a third one) next to me, said "Next stop is Brioude." AH!! As we pulled into a little jerkwater town, the conductor clearly announced: "ARVANT." I looked at the lady, she grabbed her schedule, looked at it, at the sign coming into view which said: "Arvant," shrugged HER shoulders and stared out the window!

My "first" lady came in shortly and said: "Next stop is Brioude. Bonne chance." Yeah. Next stop WAS Brioude and I disembarked, walked to the stationhouse. It was 5:45 and deserted!

This of course, set my heart racing. Was this not the right place? The right day? The right, for God's sake, country? And why wasn't I home slipping into a nice dry Martini? I flopped into a bench and decided I could do no more than wait. After ten minutes, a young, bearded man came in. He looked around, saw me and said: "Mr. Nintzel"? Yes, oh YES, that's me! Said I quite calmly.

It was Camille, of course. He led me to the car and we drove out of Brioude and into La Chappelle Laurent. A lovely, tiny French Village. At a three-story stone building encompassing many acres of land, we opened a gate and drove in. I saw many tents and cars. The cars had license plates from many places other than France.

Camille introduced me to some people, and one, George, spoke German, so we got along pretty well with broken German and dictionary

French. I was then introduced to Mme. Tu Trih, one of the "incorruptibles." These are five women Solazaref has personally trained. This lady was of Vietnamese extraction. When introduced, I said "Comment alléz vous?" She turned to Camille and laid a lot of verbs on him. Camille explained that she felt that such comments were inappropriate for this type of place. Uh oh. I explained to Camille, that in addition to that, Bon Jour was the only other French phrase I had readily committed to memory. He suggested I limit myself to Bon Jour, at least to her. I found out that this building was strictly for women and children, the wives of male alchemists who did not wish to participate in the rigors of alchemical camp life, or who just had no great interest in alchemy anyway. I met one or two and had a pleasant, if difficult, conversation. The kids, once they knew I spoke no French, had fun with me anyway. They, continually speaking French, intuiting I would divine their meaning. And, so I did, and so did they.

Camille then started to pack the little Renault station wagon, in which we had come, with food. He ran a "shuttle" three times a day from this house to the camp, bringing the appropriate meal. This one was to be dinner. And I was to have my first meal with the alchemists. We departed and passed through some lovely meadows that overlooked valleys formed by the volcanic subsidence. The view was spectacular. We passed old windmills, and finally came to a rutted road that was to lead down 1000 meters to the valley floor below.

And down we went. At the base, Camille drove to a shelter, rough, but expertly made with plastic sheets covering it all. The sheets had water in them and they acted as "air conditioners" during the day, the evaporating water keeping all remarkably cool. Half-rounds constituted benches for the tables, all sawn from local timber and lashed together. A remarkable job, I thought. Camille's arrival always, as I was to learn, brought the men, like Pavlov's dogs!! And they came, and preparations for the evening meal were underway. A tin plate, plastic bowl, cup, fork and spoon were secured for me. Soon we assembled, standing at the tables. We joined hands in prayer and the Master then arrived. He led all in a very Catholic-oriented ritual. Then we sat. I wished the half round had been placed with the flat side UP! The 'dining room attendants' started to serve. First, wine. A bottle per each four diners. Then, the customary buttered, rough bread. Then the meal.

I might as well set down here, in one fell swoop, what mealtimes were like in general, so not to repeat myself. First of all, I was amazed at how delicious everything was. And there was always plenty to eat and drink. Every morning, the wooden box of buttered peasant (like corn-rye) bread appeared. As did jars of miel (honey) and various local jams. Pitchers of cold fresh milk were passed. Boxes of "Meusli" a sort of granola was available and usually, but not always, some fresh fruit (melon, Queen Claude plums, of which I have become inordinately fond; pears, peaches, etc.). Some mornings there was cheese, usually a light bleu and always "Tomme." I brought home a wheel of this lovely fromage!

There was usually a salad with a marvelous dressing made from malt, oil, vinegar and spices. Always a vegetable. And even a simple dish of haricots vert when cooked, lovingly, with butter, garlic and spices, came out a culinary wonder. Entrees were varied and delicious. Once it was a rough pate en croute, once marinated lamb chops grilladé, once brown rice with a sort of "hash" on it. Great. One night, Nadia (an Indian—German girl) made some Indian dishes. Super. And one night, sigh, they brought out what looked like pizza pies without topping. They were actually huge shortbread "cookies" and they ladled a custard on this and covered it with the freshest, largest framboise (raspberrys) man has ever seen. Merci! Well, you have the idea now.

The first night, it was question and answer time. I was the answerer and everyone else had a 'whack' at asking questions. One chap asked, (and he was Italian — there being Italians, Belgians, Columbians, Portuguese and one Yankee) "You write people all over the world"? Yes, said I. "Have you ever written to an Italian named Luigi Veranacchi?" I said, why yes, for years, but we have never met. "I am Luigi" said he!

Another chap, from Lisbon, proved to be the best friend of the man who first told me about Solazaref! One fellow wanted to know my opinion of Western Imperialism encroaching into Europe. Say What sucker? Solazaref (yes, he was present) indicated this was an inappropriate question for this forum. Well, yeah! Solazaref announced we should celebrate my arrival by passing out some cigars; which he proceeded to do. He asked if I wanted one, as he puffed away. I said "No, I don't smoke, BUT, in honor of this auspicious occasion, I would take a puff of his!" While he roared with delight, once my statement had been translated, there was a burst of applause. Solazaref came over and asked, through the interpreter, if I thought it strange for an adept to smoke and drink. I replied in German, as I was told he spoke that language, "Not at all. In fact, I like a little toddy for my body now and then myself." He roared, ran to get a bottle of Mayville (a rather potent little drink made from tiny plums) and another burst of applause. He then asked if I thought it strange to see some of the brother alchemists wearing uniforms (they wore a sort of belted sheath). I said: "Being of German extraction, I am not only used to uniforms but put on arm bands and boots when no one is looking!"

(I wasn't sure if my subtle humor would be appreciated... but...) A TREMENDOUS burst of applause and laughter. (They loved me in Sheboygan, Mom!) And so it went until at 9:30, as per usual, we gathered around a roaring bonfire to listen to the "Mater" talk.

The first night, Solazaref allowed ANY questions to be asked. I asked about HIS Master. This was a Russian priest, Fra. Michel who came to France to teach Solazaref alchemy. After a few years, he returned to Russia to stay with the forces of Light. Last they heard, he had been put in the Gulag. (Not known is if this was by plan or simply the Soviets way of saying they didn't like Light!)

It was asked, how he came to be an adept. Solazaref responded that he was once a plain old, struggling alchemist. However, he had a visit from extraterrestrials. Following this, and a subsequent encounter, he was transformed into an "Adept." Hmmmm. He also discoursed on how positive he was that the Soviets were going, within one year, take over all of France! Now, I thought this was paranoia at its finest. I voiced an objection that this was not possible. He asked who would stop the Soviets, if they did indeed try this. I replied, "Well, the U.S." I was asked where we were when Hungary, Yugoslavia, Czechoslovakia, Bulgaria, Lithuania, etc., fell. I said, "Well this is different. France is a N.A.T.O. country." Not so, it was pointed out. There had been an assassination attempt on the life of Charles De Gaulle by a C.I.A. guy working with a NATO chap. De Gaulle felt a little 'blue' about being shot at by an ally and threw them out. So, they aren't NATO. It was pointed cut to me that there are some 10,000 KGB agents in France. They are called 'sleeping agents.' 'Sleeping' because they are keeping a low profile. Working, teaching, being regular neighbors. However, when Russia will 'make a move', they will get the word and do whatever it is they are scheduled to do (like sabotage, disrupt power, etc.).

Well, after listening to some more compelling arguments about this, I was forced to conclude that the possibility, absurd as it seems, exists. Other Europeans agreed. I mean, Russia going into Afghanistan sounds fairly absurd, doesn't it? A friend in Norway assures me that the Soviets have been harassing Sweden and Norway for some time hoping to make them take some step that can be construed as war-like and then 'retaliate' (by moving in). Remember the Russian submarine a few years ago detected in Malmö harbor in Sweden? A classic example. Turns out, the small tractor-treaded Russian subs have been detected off the coast of Japan and in the fjords of Norway! Frightening!

I was asked in various occult factions or groups in the U.S. are constantly at "war" with one another. I replied, "No. In fact, as far as I can determine, and I am certainly no expert, they pretty much cooperate with one another." This was met with surprise. Turns out French groups, be they Masons, Rosicrucians or what have you, are of the opinion if you aren't one of them, you are the enemy! I was surprised at this until Solazaref reinforced this by pointing out there had been two assassination attempts on him. The last one, by four men with machine pistols! He had been hit but had a .45 automatic with him and fired back wounding two. Solazaref is a master of Martial Arts, Zen Archery and (apparantly) the Colt .45. He kept an AK 47 in his tent at night!

Well, at 10:30 we turned in. I was assigned a tent but had brought my own sleeping bag. I slept extremely well having picked a handful of raspberries that grew in abundance as a late snack. The air was now very cool, but sweet smelling and pleasant.

At 8:00 am, a drummer beat a tattoo to wake all and sundry. The stream that ran nearby provided (icy) cold water as an aid to wake up. No one dared bathe in it. Brrrr. Few men shaved. The toilette was any space in the bushes. The "official" bathroom, and there were both ladies and gentlemen's, was a hole in the ground with a wooden shed around it to provide a modicum of modesty. The path leading to these had, as you went forward, a bush with a string attached. If the string stretched across the path, the WC was occupied. Lovely.

This day I was "shown around" and was very impressed with what I saw. First, they had found clay and built a kiln using the clay as a cover, which they baked. Then, they built a potter's wheel, from wood. They 'threw' the clay on the wheel making crucibles, retorts, dishes and the like. These were fired in the kiln. And a helluva good job at that. They built two furnaces, each one having either a pump-type bellows or holes in the structure by which, using hoses, one could puff into the thing to increase the heat.

There was an antimony mine nearby. I crawled into the entrance, donned a hard hat complete with an acetylene miner's lamp, and with hammer and chisel worked in the mine for a few hours. Quite an experience. The stream had a tributary which went through a different section of the mine and it washed out chunks of antimony. So, another task was to filter the stream searching for bits and pieces of antimony. They found plenty, high grade ore at that! The antimony was ground in a large metal mortar and pestle that Solazaref had made. (He has some devices for sale, this is one. And a thing of beauty it is!) This was "purged" (of sulphur) in one of the furnaces. In the other, iron, tartar and nitre were added to "purify" the antimony. The molten metals being poured into a metal "cone" also made by Solazaref.

We gather juniper berries, pine and "achil" (maybe a plant called 'thousand flowers'). In the morning, dew was gathered by dragging sheets on the ground. I inquired if this did not 'ground' the dew. Answer: NO. When I pointed out that Mutus Liber shows the sheet raised on stakes, I was told that the sheets in those days were often silk or other expensive cloth and they just didn't want to get them dirty. Hmmmm. The flowers and berries he gathered and mashed were covered with the dew. The idea (and here is a little alchemical "goody" for you!) is that instead of putrefying the plant (i.e., making wine), a tedious process, the dew is macerated with the plant for a week or so and then it is distilled. The first fraction coming over will be dew and... the mercury of the plant. Then, one can add more dew and using now suitable apparatus, distill off the sulphur or essential oil. This was a beautiful blue color due to the 'genveve' or juniper.

Solazaref later made a combination of the salt (laboriously calcined over a wood fire), oil and dew/mercury combination. He poured this into the herbal infusion one morning. Every morning, the hot beverage was an herbal infusion. Once in a while they had "solube" or instant coffee in addition. This, of course, was a medicine and was 'designed' to safeguard health for the camp. (And there were fifty of us!)

As you may have guessed, one of the main purposes of this 'alchemical encampment' was to teach one that even though the sources of glassware, chemicals, hot-plates, and the like, go away (as when the Russians might take over!), one can still be an alchemist and get what one needs from nature. Just like the 'ancients' did. No, not as convenient to be sure, but you can still 'operate' in a tightened environment. And we did. Oh, Solazaref also recommended rolling naked in the morning dew, which we did. This will give you energy. Actually, I needed all the sleep I could get, so I only did it one morning. And yes, there WERE women as well as men in the camp! Modesty rolled down the embankment, into the W.C.! Now, that night, even though we got drummed out of bed at 8:00 AM, I was up rapping until midnight and felt not a whit tired. Woke up bright eyed and bushy-tailed, too. Was it the dew? I think so! So, there is another little "goody" for you. Of course, this needs to be done (and sure you can wear a bathing suit when you roll on your lawn past your neighbors' window!) before the sun hits the dew. The sun will start to dissipate the energy if it shines on it.

One day the village celebrated the Feast of the Assumption of the Virgin Mary. We were given leave by Solazaref to "sit in" and had been invited by the local priests to attend mass. Armed with my trusty dictionary, I accepted a ride to town and wandered in. They had a local brass band that did a little parade down the main street of the village (La Chappelle Laurent) to the square where the proceeded to belt out some pop hits. Circa 1900! People danced in the streets, and so did I. (Some sweet young thing grabbed a hold of me. Or was it the other way around? No matter!) The smell of French cooking permeated the air and set my gastric juices bubbling. A nearby set of concessions and rides found me a "merguéz" seller. This is a North African sausage surrounded by a chunk of French bread and swathed in some snappy mustard. Then the patisserie and a Napoleon followed by an éclair (just sampling the local goodies!) At this point, I ran into Jacky Bonneau. His wife Audreé and he were at the camp. And they spoke English. I invited Jacky to the local bistro for a drink. He ordered for us, Panache. This is a mixture of a third lemonade (Seven-Up) and filled with beer. I did not think I would enjoy such a combination, but it was pretty good. Jacky asked me what I planned on doing after the camp broke up. I told him I had made a reservation at the Paris Montparnasse hotel for Tuesday and wanted to go there to meet friends. (I had written many alchemist friends in Europe to meet me there.) Also, I wanted very much to visit Pierre Munier near Lyons but did not know how to contact him. He suggested I go home with he and Audreé to their home near Rouens. Paris was less than an hour train ride away and I could easily get there when I wanted to. Since they had a car, and I would not have to travel by train for six hours, I joyfully agreed.

At the camp, Stefane Proniewski, who also spoke English, invited me to spend time with him and his wife in their Parisian apt. Unfortunately, they had to be out of town until the day I was slated to return to the States. Some other time.

That afternoon, I met with Mme. Tu Trih. She and Camille and Dominique, their publisher, talked to me about the book Solazaref had written and which was one of my reasons for coming to France.

Camille had already told me about the contents of the book. Solazaref gave techniques of becoming "in tune" with the matter the alchemist works with, how to make the vessels in the 'correct shape' so they will actually 'work' during an operation, etc. etc. I was anxious to get the book. At this meeting, it was agreed that I take the book and see if I can get it translated into English. Seems that they are as eager as I am to see it get translated. However, they were adamant in my finding the "right person," preferably someone with a 'feel' if not understanding of alchemy. This effort will probably represent the last work I do with respect to alchemical translations. If my lady, who has done so much magnificent work already, will consent to do this last item, it will be done extremely well as she is superb. She is a master of French and German and is quite "in tune" with alchemy due to certain personal experiences. So, I trundled "La Introitus Lapide Philosophorum" home with me and will soon start to see if in fact I can arrange to get this tome translated. Now, the Filiation also suggested two other items, one by Canseliet and one by Atoréne, to be translated, but these will be of lesser priority than Solazaref's book.

On the last day, it was clean-up time. Everything was dismantled, disassembled and everything restored to the condition as when we arrived. The idea was, to let it look like no one had been here! Furnaces, kiln, wheel, etc., all were taken down. What was done with the heavy logs used to build the eating place, I am not sure. But, when done, it was very difficult to tell anyone had been here.

Following the noon-day meal of haricot vert, coté de agneau and fresh framboise and cream, Solazaref took a sort of medallion. He cut it into four parts. Gave one fourth to the Italians, one to the Portugese (Dr. Estefan de la Miranda with whom I had become very friendly), one to the French and one to me as a mark of brotherhood and fraternité. He told me, this medal was transmuted from tin to silver. And, if I wished, I could have it tested. Hmmmm. In any event, a nice gesture. He told me he would like to come to Texas to join in the Sweetwater rattlesnake round-up. I told him to "come on, he could stay with me and we could transmute a little more tin into silver." He bellowed with approval. He has a lovely sense of humor! Finally, he may just DO that as he would like to consider starting an American branch of the Filiation. I told him to holler when he was ready.

We made our farewells and Audreé and I started walking up the hill. Jacky's Renault was a little puny and he didn't want to go up the rutted road with all the people and all the suitcases. So, we left the suitcases and started the 1000 meter hike. The air is thin there, folks. In less than 100 meters I was huffing and puffing. Audreé, bless her heart, suggested we stop to catch our breath. About five or six of these and we were close to the summit. And it was also quite hot. Just as I thought I would drop from exhaustion, Jacky came wheeling by and picked us up. Whew.

It is about a six-hour drive to Buchy where the Bonneaus lived. So, I was treated to more French countryside views. In Southern France, most of the buildings have red tile roofs. Just like L.A.! (or Spain, I suppose.) Now, the French like to drive at 130km/hour about 2 feet from the person in front! Outside Clermont-Ferrand, we ran into a serious traffic problem. Seems a cyclone had struck an hour before us. We found it killed 54 people, wiped out a trailer camp and caused considerable structural damage. Much of which we saw. So, we got delayed for a fair amount of time. Then freeway construction in Paris caused further delays. One time, on a hill, Jacky had to hit the break in an emergency stop. We slid ten feet sideways. I was unconcerned, merely requesting some toilet paper! I told Jacky and Audreé and Pierre (from Burxelles) that I was anxious to have a large bowl of ice and some grape soda or orange soda. I explained that this was Jus de Raisin with L'eau avec bullé. (Grape juice with sparkling water). You know, fruit flavored soda water. They asked, a little incredulous, if I really liked that sort of thing. "You bet," said I. "With lots of ice." Ice, you see, is very rare in the smaller villages. Well, we stopped to eat about eleven PM at a country inn. I went to the W.C. and when I returned, on the table was a bottle of Jus de Raisin and a 'fifth' of Limonade (like carbonated water with a squeeze of lemon). And, a bowl of ice. I was quite touched by their thoughtfulness. And drank it all.

Dinner was Prix Fixe at 70 francs. (about 11 dollars) We had salad, melon in port (for me), canard fume for the Bonneaus. I had Coq au Vi, they had escargot (many), for dessert I had poivre belle Helene. (Ice cream and chocolate sauce over a poached pear. Yummy.) Then the obligatory plate of various cheeses. A fluffy goat cheese was my favorite although I managed a little of the Tomme and a little of the Brie. Some excellent Pouillée Fuse (we were in that region) washed it down. Not bad for 11 bucks. However, everything else, except for this Prix Fixe dinner, is fairly expensive.

The Bonneaus live in a house that is 250 years old. Across the street, the farmer kept cows. I introduced myself, the next morning on a walk, to the cows and offered to conduct a Tupperware party for the ladies that night! They stared mutely at me. Sigh!

When we got to their house, I HAD to take a shower. We're talking lots of brown water going down the drain! And a night in a bed with clean sheets. And no drummer in the morning. Ah. Audreé made some wonderful croissants, some confiture, beurre and miel and hot coffee. Yeah man. This camping jazz is alright, but I find I am a creature of comforts, by-and-large.

We spent the day calling the Sheraton-Montparnasse to see if I had any messages. NONE. Seems one chap from Paris (as indicated, I wrote many European friends to try and meet me in Paris) was going to the States as I was going to France. Ha. Also, in Spain, Julian and Manuel had set out on a tour of Europe themselves. Urs in Switzerland was busy, Jose in Portugal was in Israel, Jean in France was on vacation (as were most Frenchmen during the month of August) and so on. Well, I managed to find Pierre Munier in Lyons who could not make it to

146

Paris, nor I (now) to Lyons. He actually lives in India and we had never talked or met, just corresponded. Now we had a nice, satisfying chat on the telephone. In fact, he and the Bonneaus got to talking. With that, I cancelled my hotel in Paris and opted to stay with my new-found friends, Jacky and Audreé, for another day.

And it was a delightful time. We talked about alchemy, for the most part, and life in France vs. the U.S., life in general, and of course, books! Jacky has a marvelous library. He gave me a couple of books, in English, that he also has copies of in French. I gave him my personal copy of "Golden Chain of Homer" and he gave me a book of the color plates of "Splendor Solis". What generous, loving people they are. The next day, they drove me to Rouens. We visited the incredible Cathedral. Astonishing. The markets, shops, a museum or two and then back home where Audreé had made, for me, some wonderful soup along with the Pate and other goodies. The next day, she gave me a bag of Queen Claude plums, which I so loved, and put a large rose from her garden in it. I told her roses were my favorite flowers.

The next day, after a tearful farewell, Jacky drove me to the train station. I got off at Gare d'Lazare and grabbed a cab. He spoke good Italian, but little English. However, he took me under the Arc d'Trimophe, down the Champs Elysee and to Notre Dame Cathedral. It was pretty whirlwind, alright, but I did get to see some of Paris. It is wonderful, but you know what? I will take the U. S. of A. anytime. We are just not aware of what we have in our own backyard. Yes, I loved visiting France. As I did England. Would go back at the drop of a hat. And mostly because of the many friends I made there!

In sum, it was a magical tour. Many things, I feel, were 'started' and remain to gestate before they come to fruition. What are they? I don't

147

really know. I can only feel that there is SOME thing. I learned some things, got different viewpoints on many things and, most importantly, met some wonderful people along the way who are stars in my galaxy of marvelous friends.

As a sort of postscript, when I got back, I started my new job with Kentek Services, Jody had to go into the hospital for two days for an Angiogram. She has blockages, the bad news. However, the good news is they think they can alleviate some of this via oral medication. We will just have to see.

Shortly after, I virtually collapsed. I was 'down' all Labor Day weekend, running 102+ fever. The next weekend, same thing. What? Not sure, but we figure it was no doubt stress-related. I have been "going" for six months. Up every night for months until 2:00 am writing and working translated items, corresponding, etc. Also, trying to keep the business going, various other problems that were quite stressful, etc. Then when Jody announced she had to have the Angiogram, well, that might have been the proverbial 'last straw' and my body caved in.

We are both in fine shape as I write this. Jody is taking care of herself and I have been going to bed at 11:00 and getting up at 6:30. Even on weekends. So, some physical changes have taken place, in addition to some mental and spiritual ones! What all this is leading to, only time will tell. Anyway, there it is, a bit lengthy chronicle of my Alchemical Mystery Tour to France. I hope you enjoyed my recounting of it as much as I have in writing of it. PEACE and LOVE.

In L.V.X.

Hans

Philip N. Wheeler

A Word from the Publisher

Thank you for purchasing this book from The R.A.M.S. Library of Alchemy. During his lifetime, Hans Nintzel dedicated himself to the identification, acquisition, study, retyping and, when necessary, translation of what he considered to be the most important known works on Alchemy. Hans was assisted by his sparse network of fellow Alchemists, all members of the Restorers of Alchemical Manuscripts Society (R.A.M.S.). I was an active member of R.A.M.S.

Hans provided copies of the R.A.M.S. works as photocopies. My goal is to publish all of them as professionally printed books.

The works from the original R.A.M.S. Library are republished by R.A.M.S. Publishing Company in the collection, "The R.A.M.S. Library of Alchemy," with permission of the Estate of Hans W. Nintzel.

If you have a work on Alchemy that you believe should be a part of the R.A.M.S. Library, please contact me through R.A.M.S. Publishing Company.

Philip N. Wheeler

https://ramsalchemy.jimdo.com/

Philip N. Wheeler

The R.A.M.S. Library of Alchemy

The study and practice of Alchemy was extremely important to Hans W. Nintzel. He assembled this Library over a period spanning more than three decades, guided by his teacher Frater Albertus. The R.A.M.S. Library of Alchemy includes all of the most valuable Alchemical texts that Hans painstakingly located, acquired, retyped, and translated during his lifetime, with help from other R.A.M.S. members.

The following is a list of the volumes that are currently available. Volumes that contain works from multiple authors may have only the principle author or editor listed.

Volume	Title	Author or Editor
1	Twelve Keys of Basilius Valentinus	Basilius Valentinus
2	Triumphal Chariot of Antimony	Basilius Valentinus
3	His Secret Book	Artephius
4	The Golden Work	Hermes Trismegistus
5	Three Works of Ripley	George Ripley
6	Four Works of Paracelsus	Paracelsus
7	Bacstrom's Notebooks, Part 1	Sigismund Bacstrom
8	Bacstrom's Notebooks, Part 2	Sigismund Bacstrom
9	Summa Perfectionis	Geber (Abu Musa Jabir ibn Hayyan)
10	The Five Centuries	Rudolph Glauber
11	The Greater and Lesser Edifyer	Johann Grashoff
12	Chemical Secrets and Experiments	Sir Kenelm Digby
13	The Turba Philosophorum	Arisleus
14	Das Aceton	Christian Becker
15	The Art of Distillation	John French
16	Non-Violent Destruction of the Atom	Nintzel & Wheeler
17	Philosophical Furnaces	Rudolph Glauber
18	The Last Will and Testament	Basilius Valentinus

19	TBD	
20	TBD	
21	Alchemical Symbols, Fourth Edition	Philip N. Wheeler
22	The Book of Formulas	John Hazelrigg
23	18 Short Tracts	Hans W. Nintzel
24	Bacstrom's Notebooks, Part 3	Sigismund Bacstrom
25	A Discourse on Fire and Salt	Blaise Vignere
26	The Mineral Work	Johan Hollandus
27	The Vegetable Work	Johan Hollandus
28	Lamspring's Process	Lamspring
29	The Book of Abraham the Jew	Abraham Eleazar
30	Five Short Works of Glauber	Johann Glauber
31	The Metamorphosis of the Planets	Johannes Monte-Snyder
32	Four Works of Roger Bacon	Roger Bacon
33	The Golden Chain of Homer	Homerus, Kirchweger, Nintzel, Wheeler
34	Alchemy Rediscovered and Restored	Archibald Cochren
35	Aurifontina Chymica	John Houpreght
36	The Golden Fleece	Salomon Trismosin
37	The Transmutation of Base Metals into Gold and Silver	David Beuther
38	Sanguis Naturae	Christopher Grummet
39	A Revelation of the Secret Spirit	Giovanni Lambi
40	The Holy Guide, Part 1	John Heydon
41	The Holy Guide, Part 2	John Heydon
42	Secreta Alchymiae	Kalid Persica
43	The Golden Treatise of Hermes	Hermes Trismegistus
44	Potpourri of Alchemy, Part 1	Hans W. Nintzel
44	Potpourri of Alchemy, Part 2	Hans W. Nintzel
46	TBD	
47	Selected Chemical Universal and Particular Processes	Alexius von Ruesenstein

http://ramsalchemy.jimdo.com

www.ingramcontent.com/pod-product-compliance
Lightning Source LLC
Chambersburg PA
CBHW081726220526
45468CB00008B/1989